HAMSTERS

HAMSTERS
The Ultimate Pocket Pet

Virginia Parker Guidry

**photographs by
Carolyn McKeone**

·PRESS·

Irvine, California

I-5 PUBLISHING, LLC™
Chief Executive Officer: Mark Harris
Chief Financial Officer: Nicole Fabian
Vice President, Chief Content Officer: June Kikuchi
General Manager, I-5 Press: Christopher Reggio
Editorial Director, I-5 Press: Andrew DePrisco
Art Director, I-5 Press: Mary Ann Kahn
Digital General Manager: Melissa Kauffman
Multimedia Production Director: Laurie Panaggio
Multimedia Production Manager: Jessica Jaensch
Marketing Director: Lisa MacDonald

Photographs copyright © 2004 by Carolyn McKeone
The hamster models in this book are courtesy of: Jordon Campbell, pp. 3, 19, 20, 21, 31, 37, 48 (top and bottom), 59, 65, 67, 71, 79, 89, 92, 95, 108 (bottom); Pet Paradise, London, Ontario, pp. 2, 6, 16, 27 (bottom), 30, 33, 34, 40, 41, 45, 73, 78, 81 (top and bottom), 87, 90, 91, 99, 100 (top and bottom), 108 (top), 112; Super Pet, London, Ontario, pp. 9, 10, 11, 12, 15, 17, 22, 25, 27 (top), 38, 42, 53, 57, 61, 62, 64, 77, 85, 88, 102, 103, 105, 115; and Jillian Mundt, pp. 54, 55, 70, 75, 106.
Cover photo; Anna Subbotina/Shutterstock

The Library of Congress has cataloged an earlier printing as follows:

Guidry, Virginia Parker.
  Complete care made easy. Hamsters : the ultimate pocket pet / by Virginia Parker Guidry ; photographs by Carolyn McKeone.
    p. cm.
  Includes index.
  ISBN: 1-931993-31-9 (softcover : alk. paper)
  ISBN-13: 978-1-931993-31-9
  1. Hamsters as pets. I. Title: Hamsters. II. McKeone, Carolyn. III. Title.

SF459.H3G85 2005
636.935'6—dc22

                    2004003349

I-5 Press™
A Division of I-5 Publishing, LLC
3 Burroughs
Irvine, California 92618

Printed and bound in China
13 12 11   6 7 8 9 10

Acknowledgments

WHILE WRITING IS A SOLITARY TASK, NO BOOK IS WRITTEN without the help of others. This book is no exception. So, hats off and many thanks to the hamster enthusiasts who answered questions, made suggestions, and clarified information. And, thank you to Dr. Max of long ago, who delighted a youngster with imaginative hamster stories.

—V.P.G.

FOR PAMELA HANNANT, MAY YOUR LIFE BE ENRICHED by all living creatures. Also, in memory of my dear husband Pete, who encouraged me to return to photographing animals and birds.

—Carolyn

CONTENTS

Foreword

One of the first things I learned about hamsters is there is a lot more to them than meets the eye. Did you know there are more than twenty hamster species? Don't worry, you don't have to learn about all of them as only a few are kept as pets, but each species is unique. This book makes learning about these fascinating pets fun—it's easy to read and the answers to your questions are easy to find. The more you read, the more you'll discover why the hamster is one of the world's most popular pets.

My sister, Andrea, adopted a golden hamster when she was in the 10th grade. Looking back, she remembers she didn't give the hamster the attention it needed, and especially didn't clean the cage enough. In college, her boyfriend (now her husband) got her a hamster named Lucy. Unfortunately, Lucy was pregnant, and instead of having one pet, my sister found herself with multiple hamsters. The original pet store refused to take back the babies, but she did finally find a store that would take them.

Experiences like my sister's can be typical for small animal owners. Although we all must learn some things through experience, it benefits both you and your pet to begin your relationship on the right foot—armed with the knowledge of proper hamster care. So whether you've just adopted an adorable and fun-loving hamster who will give you hours and hours of enjoyment and entertainment or are just thinking about getting one, *Hamsters: The Ultimate Pocket Pet* can help you make sure your new pet thrives and not just survives. Read on, and get ready for a wonderful and rewarding friendship!

—*Melissa L. Kauffman*, Digital General Manager, I-5 Publishing, LLC

CHAPTER ONE

Hamster Defined

Hamsters. Some call them the world's most popular small pet. To be sure, hamsters are a much-loved and much-enjoyed pet in many countries. They are the primary subject of many books, newsletters, countless Internet Web sites, and even a television show. Because of hamsters, clubs are founded, children are introduced to the world of pets, and companies manufacture an array of specialty products. Amazing, isn't it?

Yes and no: It is amazing that such a tiny critter has such a huge influence. But when you learn more about hamsters, you begin to understand the almost magical spell that they have cast upon their adoring fans.

Rodentia

The hamster is a member of the largest group of mammals, called *Rodentia*. The *Rodentia* order also includes beavers, musk-rats, porcupines, squirrels, prairie dogs, and many other small mammals. With more than two thousand living species catego-rized in about thirty different families, rodents make up over 40 percent of mammal species. Hamsters belong to the subfamily *muridae*, which consists of more than 1,100 species including rats, mice, voles, muskrats, lemmings, and gerbils. Within the *muridae* family, there are more than twenty species of hamsters. Only a few are kept as pets, though. The others live in the wild throughout the world in areas as diverse as Africa, Asia, and Europe. The *Rodentia* order does not include rabbits, as many people mistakenly believe. Rabbits and hares belong to the *Lagomorpha* order.

Hamsters belong to the same family as rats, mice, and gerbils.

Rodents have extremely diverse lifestyles. Some species live in rain forests, others live underground, and still others make their homes in deserts. They range in size from pygmy mice weighing .05 kilogram to capybaras weighing some 70 kilograms.

Like all rodents, a hamster has teeth made especially for gnawing. All rodents have a single pair of upper and a single pair of lower incisors, followed by a gap (diastema), and then one or more pairs of molars or premolars. No rodent has more than one incisor in each quadrant, and no rodent has canine teeth. Rodent incisors are rootless, which means they grow continuously. Rodents gnaw with their incisors by pushing the lower jaw forward and chew with the molars by pulling the lower jaw backward. As a rodent gnaws, the incisors grind against each other and wear down the soft dentine. This natural wearing keeps the teeth sharp and at a proper length. To accommodate their chewing patterns, rodents have extremely strong jaw muscles.

Hamster History

The modern story of the hamster began in 1829, when a British zoologist named George Waterhouse discovered the small rodent near Aleppo, a city in Syria. Waterhouse promptly named the creature *Cricetus auratus*, meaning "golden hamster." The hamster enjoyed a brief period of popularity in England, but interest soon waned.

Then in 1930, Professor Israel Aharoni, a zoologist at the University of Jerusalem, traveled to Syria to investigate the "Syrian mice" that children in the area reportedly kept as pets. The professor's investigation was fruitful. He discovered a mother and litter of these "mice" in an underground burrow and promptly took them back to Jerusalem for study. He named the creatures *mesocricetus auratus*. It is believed that some of the captured hamsters escaped or died, and Dr. Aharoni was left with

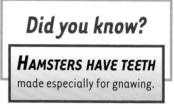

Did you know?

HAMSTERS HAVE TEETH made especially for gnawing.

two females and one male. The remaining hamsters mated, and within a year, they produced numerous offspring.

In 1931, one year after discovering them, the professor sent some of the animals to England for further study and breeding. Then in 1938, offspring were sent to the United States for research and eventually ended up in the hands of private individuals who had an interest in breeding them. By the 1950s, hamsters were the rage among pet lovers in the United States.

And they still are! Walk into any pet supply store, and you're certain to find a hamster or two on display or for sale, as well as a host of hamster gear. Hamsters can be found in classrooms, in children's bedrooms, on exhibition at shows, and as the topic of both fiction and nonfiction books.

Why the fuss over such a small creature? Because the hamster is the ultimate pocket pet! Small, cute, fuzzy, clean, and easy to care for are just a few of the hamster's star qualities. Hamsters are great beginner pets for children and excellent pets for adults. They're fun to watch, inexpensive to keep, and easy to tame.

Of course, if you've just purchased a hamster, you probably already know this. No extra convincing of the hamster's merits is necessary. What is necessary is information on how to correctly care for your small pet. The following is just that—and a whole lot more!

There are more than twenty species of hamsters.

CHAPTER TWO

The Right Choice

There's no denying the hamster's appeal as a pet. These small, cute, and clean companion animals are easy to care for and fun to watch. Once you choose to own a hamster, there's no turning back. You're hooked!

But owning a pet starts with a choice. Pet ownership doesn't just happen; you choose to become a responsible and kind owner. You may think that because hamsters are so small and low maintenance, you don't need to give owning one a second thought. No dramatics intended here, but *give* it a second thought.

Great Owners Only!

Hamsters—as well as all companion animals—deserve responsible owners who think through owning a pet. Ask yourself some questions: Is a hamster the right pet for me and my family? Do I have time to learn about hamsters and their care needs? Can I devote the time necessary to keep this small fuzzy friend well fed, safe, and healthy? Will I be able to provide care for the hamster's lifespan? Can I make an investment of time and resources to this tiny creature, who will be totally dependent upon me and may even keep me awake at night? (Yes, you read correctly. Hamsters are nocturnal animals, being most active in the early evening and throughout the night.) Regardless of a companion animal's size, it will make an impact on your life. It's true that a pet hamster makes a substantially smaller impact than a large dog, which requires intensive training, a large yard, and 50 pounds of food a week. But it makes an impact nonetheless, and you might need to make some adjustments to accommodate this tiny friend. Be aware of that, and be ready to make changes.

If you're not willing to adjust your lifestyle—even a little—then perhaps a pet hamster, or any pet, isn't right for you. That's okay. Better to know that before you bring home a hamster than after. Unfortunately, the latter happens all too often. A person impulsively decides to bring home a pet and then realizes how much care is involved and wishes the pet would go away. That's how many animals end up in shelters and foster care without owners to love and tend to them. You can avoid this scenario by thinking through your idea of owning a hamster first. Then make a decision to be the best owner you can be.

Child's First Pet

Hamsters are common first-time pets for kids because they're small, easy to care for, entertaining, and inexpensive. Even kids who don't have a hamster at home will probably be introduced to

Hamsters are nocturnal creatures.

one while attending preschool or kindergarten. Owning a hamster is a wonderful first opportunity for a child to get acquainted with an animal, and that child usually becomes very attached to the hamster. That special attachment is a good way for a child to begin learning about friendship, empathy, and selfless caring.

The friendship between a hamster and child must be supervised by an adult. Even the most responsible, mature child eventually does something irresponsible—not because he's lacking in character but because he's not yet mature. Children are not born knowing how to relate to animals, and the young owner of a hamster is no exception. Parents must teach children how to take care of and handle hamsters.

The best way for a parent to teach a child how to handle a hamster appropriately is to model correct behavior. If the child

Show children how to hold a hamster properly.

sees the parent holding the hamster gently, he learns to do the same. You cannot expect a child to handle a hamster correctly if you don't show him the right way to do it. When demonstrating, tell the child what you are doing (touching softly) and why (so we don't hurt the hamster). Make it understandable to the child by breaking the desired behavior into small steps. Be positive and praise the child when he handles the hamster correctly.

One mistake parents frequently make is assuming their young children are capable of taking care of a pet all by themselves. They aren't. Parents are ultimately responsible for household pets, although children, depending upon their age and maturity, can take charge of various duties. Young children can do simple tasks like filling the hamster's feed dish; older children

Golden hamsters are the most common pet hamsters.

can clean the cage. However, parents must always make sure the jobs are done correctly.

The Lineup: Hamster Types

Of the more than twenty species of hamsters, only a few are kept as pets. The following are the most common pet hamsters.

Golden hamsters, also known as Syrian hamsters (*Mesocricetus auratus*), are a direct descendent of the creatures captured by Israel Aharoni. Today's golden hamsters are probably the most common and popular pets. They come in a variety of colors, such as black, cinnamon, golden, sable, and gray. They also have various coat styles, such as shorthair, longhair, satin, and rex. Other names for golden hamsters

Dwarf hamsters can squeeze through the bars of some hamster cages.

include "standard," "fancy," and "teddy bear," the latter of which has long hair. Golden hamsters are about 7 inches long with hairless feet and a small tail. Golden hamsters are fairly good-natured and make a great pet. They can live from eighteen months to three years.

Originating in the desert lands of the Middle East, golden hamsters are solitary animals. Therefore, in the wild, each has its own burrow and surrounding territory. The male golden hamster is lured to the female's burrow only when she is in heat. After mating, the female remains alone in her burrow, foraging for food at night to add to her cache so there will be sufficient nourishment for both her and the newborn pups. She raises the pups alone and bans them from the burrow as soon as they are weaned.

Smaller than golden hamsters, the dwarf Campbell's Russian hamsters (i Phodopus campbelli) were introduced to the pet market in the United Kingdom in the 1970s. They are seen in pet shops around the world and are considered good pets, although they can be less tolerant of clumsy handling and may nip. Dwarf Campbell's Russian hamsters live one to two years. Solid glass or plastic cages are recommended because these hamsters can squeeze through the bars of some hamster cages. The original color of wild dwarf Campbell's Russian hamsters are brownish gray, but today these hamsters are available in a variety of coat colors.

Wild dwarf Campbell's are found in the deserts and sand dunes of Mongolia, where winters are extremely harsh. These dwarf hamsters are able to withstand the cold because they have a thick coat covering their entire bodies, including the feet. The fur on their feet gave them the name "furry-footed hamster" when they were first domesticated. Campbell's live in pairs and normally inhabit the abandoned burrows of other rodents. They remodel the burrows by adding additional escape and ventilation tunnels, with the original nest remaining as the primary living quarters. Just prior to giving birth, the female builds an additional nest for the expected litter. Because the pups are unable to maintain their body temperature for the first twelve days, it is imperative that either the male or the female remain on the nest with them to keep them warm.

The diminutive dwarf winter white Russian hamsters (i *Phodopus sungorus*) were introduced to the United Kingdom pet market in 1978 from southwest Siberia. Also called Siberian hamsters, dwarf winter white Russian hamsters are so named because their gray coat turns white in the winter months. These hamsters are not as common a pet as the dwarf Campbell's Russian hamsters. Because of their small size, these hamsters aren't gen-

Dwarf winter white Russian hamsters are found primarily in Siberia.

erally recommended as children's pets, though they are generally good-natured. Dwarf winter white Russian hamsters do not live as long as golden hamsters; their life span is usually one to two years.

In the wild, dwarf winter white Russian hamsters are found primarily in Siberia. However, some are found in the same territory as the Campbell's—in the deserts and sand dunes of Mongolia. They are also able to withstand extremely cold temperatures by residing in pockets of ungrazed land on flat steppes. During the winter months, they have a natural camouflage, as their coats turn all white, with the exception of a dark gray-brown dorsal stripe and two flank stripes. Unlike the Campbell's, who live in pairs, the dwarf winter white Russian female normally shares her burrow with at least two males. The males, in return, usually share burrows with at least two females. Each pair has its own burrow and also shares a different burrow with another mate. When the female is in heat, she will mate with both males. As with the Campbell's, the males assist with caring for the young.

The mouselike Chinese dwarf hamsters (*Cricetulus griseus*) come from northern China and Mongolia and have been used in laboratories in the United Kingdom since 1919. Chinese dwarf hamsters were introduced to the pet market in the 1970s. Though not common, Chinese dwarf hamsters make good pets. They are brownish gray with a black stripe on the spine and a white tummy. They can be timid and fast moving—difficult to catch if lost. Chinese dwarf

> ## Did you know?
>
> **HAMSTERS PROCRE-ATE** extremely quickly. Gestation averages sixteen days. It is best to house hamsters separately to avoid unwanted litters.

This Chinese dwarf hamster plays with his toys.

Hamsters love to play in nests they can burrow in.

Hamster Stats

- Golden hamster (6–7.75 inches long)

- Dwarf winter white Russian hamster (3–4 inches long)

- Dwarf Campbell's Russian hamster (4–7.75 inches long)

- Chinese dwarf hamster (4.25–7.75 inches long)

- Roborovski hamster (1.5–2 inches long)

hamsters usually live two to three years. Because of their small size, Chinese dwarf hamsters should be housed in solid glass or plastic cages.

Chinese dwarf hamsters are found naturally in extremely rocky country. They live together in pairs and make their burrows out of natural crevices in the rocks with two entrances. A pair may have a litter every 21 days, and the pups stay with the mom until they are fully weaned at three weeks.

The least common hamster pets are probably the sandy brown Roborovski hamsters (i Phodopus roborovskii), which come from Mongolia and Northern China. These dwarf hamsters were imported to the United Kingdom by the London Zoo in the 1960s. These hamsters are lively and fast and are therefore known to be difficult to handle. They are not usually seen in pet stores, though they can be purchased from breeders. Solid plastic or glass housing is recommended, as is a large cage to accommodate their high activity level. Roborovski hamsters usually live up to three years.

Roborovski hamsters naturally live in an area with flat sandy soil. Their normal breeding cycle is during June and July, when two or three litters are born in succession. In most cases, the pups won't begin producing until the next breeding season. Because their burrows are built close to the surface, they are more vulnerable to predators. But nature has endowed them with longer legs, allowing them greater speed for an escape. They are excellent climbers and use their long tails to assist them in gripping the rocks. Because they live in a warmer climate, they are not as well adapted to colder regions as the other dwarfs.

Small or Extra Small?

For years the golden hamster has been the main staple of the pet hamster market. Increasingly, dwarf hamsters, including the Campbell's Russian, winter white Russian, and Chinese, have become popular with hamster enthusiasts. According to those who know and love these tiny creatures, the dwarf species tend to be more sociable with each other.

This makes sense, given that in the wild the dwarf species of hamsters mostly live in pairs, unlike the solitary golden hamster. The breeding season of the dwarfs depends upon the climate in the territory of their residence, but because of the cold winters, most of them raise their young in the spring and summer months. However, some enthusiasts say that while the dwarf may have social skills with their own kind, they tend to be more prone than the golden hamster to nipping their keepers.

CHAPTER THREE

Locating a Healthy Hamster

IF OWNING A HAMSTER IS EXACTLY YOUR CUP OF TEA, you need to know where to find one. Fortunately, that's not difficult. You can easily purchase a hamster at a pet store or from a breeder, or you can adopt one from a shelter.

Pet Store

Perhaps the easiest and most popular place to purchase a hamster is a pet store, where you're sure to find hamsters displayed in cages. Pet stores usually have hamster gear—such as cages, toys, food, and squeaky wheels—on hand, too, which makes stores a convenient place to buy.

When it comes to hamsters, not all pet stores are created equal. Buy only from a store that takes excellent care of its animals. The hamsters should be kept in appropriate housing that is clean, dry, and relatively odor free. The hamster should be healthy, with no signs of illness. Healthy hamsters are alert and curious and have attractive coats, clean ears, and bright eyes. The staff should be knowledgeable and able to answer questions about individual species for sale.

Breeders

Another good way to locate a hamster is to contact a breeder, especially if you are looking for hard-to-find varieties such as the Roborovski hamster. Breeders are individuals who breed and sell hamsters. As with pet stores, not all breeders are created equal. You want to locate a responsible breeder who breeds hamsters carefully, with consideration of the offsprings' health and temperament. A responsible breeder knows hamsters and how to care for them properly.

To locate a breeder in your area, ask for a referral from a local pet store or hamster club (see appendix on page 111). It is also a good idea to ask your veterinarian or a friend who owns a hamster for the name of a reputable breeder.

Helping Hamsters

Did you know that hamsters are routinely abandoned by their owners and left homeless? Hard to believe, but it's true. There are many reasons owners give up hamsters—they are moving, the kids are bored, the hamster bites. There are other reasons, of course, but these are common ones.

What happens to unwanted hamsters? Many are given away, let loose in the neighborhood, or returned to the pet store. And, increasingly, unwanted hamsters end up at animal shelters

Hamsters should be kept in appropriate housing that is clean, dry, and odor free.

and rescue organizations. Unfortunately, many would-be hamster owners aren't aware of the many hamsters awaiting homes at these organizations.

If you're seeking a pet hamster, consider visiting your local animal shelter. They're a great source for adopting a pet hamster and provide a great opportunity for you to rescue a small and needy creature. Be sure to ask if there are hamsters available; these small creatures aren't always placed on display. If there aren't any hamsters available, leave your name and ask for a call if one is relinquished.

Also check with organizations that specialize in rescuing hamsters. Rescue organizations are usually staffed by dedicated volunteers who would love to place a hamster in a good home. You can find one near you by checking online or by looking in the yellow pages under "animal shelters" or "humane societies" to find shelters in your area. Give the organization a call to find out if hamsters are available. Another possible source is a friend or neighbor who happens to have a hamster with a litter. Most people are happy to find good homes for their baby hamsters—for free.

What to Look For

Once you've decided where to obtain a hamster, your next step is to choose a healthy pet. How can you tell if the hamster you're thinking of taking home is healthy? First, consider the hamster's overall appearance and behavior. You want a hamster that is active (though daytime sleeping is normal), curious, and well proportioned from head to tail and that has soft, clean fur with no bald spots, clean ears, and bright eyes. Healthy hamsters are busy scurrying about their home, eating, drinking, and running through tunnels or on a wheel.

Healthy hamsters have clean ears, bright eyes, and no bald spots.

Next, take a good look at the hamster's environment. A good environment is essential because it reduces stress and helps the hamster stay healthy. Is the cage clean and dry? Is it uncrowded? Is it equipped with appropriate food dishes, toys, and a water bottle? Are there any signs of illness?

There are several telltale signs of ill health of which you must be aware. Avoid buying a hamster that exhibits any of these symptoms:

- diarrhea or a wet rear end (could indicate a highly contagious illness commonly called wet tail)
- hair loss or a coat that is rough and dull
- lethargy
- low body weight
- runniness or discharge from the eyes

When choosing a hamster, it's important to consider age as well. Unfortunately, the adorable hamster is not blessed with a long life. Most hamsters live two to three years. Hamster enthusiasts often recommend buying a baby about eight weeks old and raising it yourself rather than buying an adult. That way, you'll have the maximum amount of time with your small pet. Also, you can raise a young hamster to accept handling at a young age. However, there's no harm in bringing home an older hamster. Treated correctly, the adult hamster can bond with his new owners, though it may take time.

Do you want a male or female hamster? Hamster enthusiasts say there's not much difference in behavior and that either sex makes a good pet, so it's really your preference. Females are generally a little larger than males. Females are usually between 95 and 125 grams, whereas males are between 85 and 120 grams. You can determine the hamster's sex by inspecting the rear end.

The distance between the anus and genitalia is greater in the male than in the female. To prevent unplanned pregnancies and fights, remember to house hamsters separately if you're planning on bringing home more than one. Also, if you're bringing home a female, try to make sure she isn't pregnant by asking if the hamster was kept alone. Check to see if she is more round than other females unless you want a ready-made family.

After taking into account the hamster's health, age, and sex, the best advice on choosing one is to pick one that appeals to you. Is there a

Did you know?

OBEY THE GOLDEN RULE of hamster ownership: Never wake a sleeping hamster.

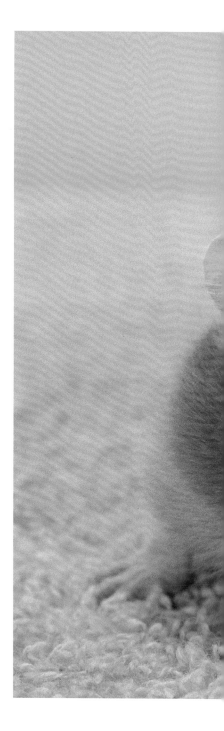

Choose your new pet in the late afternoon when hamsters are most likely to be awake.

particular hamster that seems friendly and outgoing toward you? Don't be in a hurry. Take time to watch the hamsters carefully. You can learn a lot about a hamster's health and personality just by watching.

Additionally, because hamsters are nocturnal, enthusiasts recommend visiting with and choosing new pets in the late afternoon when the hamsters are most likely to be awake. You can tell little about a small animal curled up sleeping soundly. However, waking a sleeping hamster is a no-no. Hamsters tend to be grouchy when awakened. You might get nipped.

Won't My Hamster Be Lonely?

The general rule with hamsters, specifically golden hamsters, is one hamster per cage. Why not house more than one per cage? Aren't hamsters lonely? Golden hamsters are solitary animals.

Dwarf hamsters are the exception to the one-hamster-per-home rule.

Those Cheeks!

IT WON'T TAKE LONG BEFORE YOU FIGURE OUT THAT hamsters have something different about their cheeks: that is, they're packed with breakfast, lunch, and dinner! Hamsters have cheek pouches that expand as they forage for food. This allows them to store food that isn't eaten right away and carry it back home to their burrows, where they empty the pouches by stroking their cheeks with their paws. Hamsters with full cheek pouches are quite a sight, one hamster enthusiasts love to see.

They aren't lonely when kept by themselves. By instinct, they live alone in burrows. Once the male mates with the female, he is off to his own burrow. He takes no part in raising young, and the female doesn't tolerate a male near her young. When golden hamsters are four to five weeks of age, they should be taken from their mother and given private individual quarters. If you fail to do this, chances are fights will break out—and the squabbles can be nasty, resulting in severe injury or even death. To keep your hamster happy and stress free, you must respect her solitary nature.

The exception to this one-hamster-per-home rule is with dwarf hamsters. In some cases, dwarf hamsters can be housed together. They are naturally more sociable, and pairs mate for life. The males aid in rearing the youngsters. Two females or two males can sometimes be housed together peacefully if they have been raised together.

CHAPTER FOUR

Welcome Home

Have you ever thrown a welcome home party for a friend or relative? There are plenty of preparations to be done, aren't there? In fact, planning can take weeks. You've got to plan where to have the party, what type of party, what time, what's on the menu, how many guests, and so on. Before you bring home your new hamster and welcome her with a coming home "party," you must prepare. You must think about what you'll need to care for your small furry friend. There are a number of decisions to make and supplies to buy.

Don't think you need to plan ahead for your hamster's homecoming? Well, the idea behind planning is to make life easier for all—the hamster, you, and your family. If you're prepared, it's likely to be a stress-free transition. Plan ahead!

There are some basic supplies you'll need to care for your hamster. Hamster gear can be purchased at almost any pet supply store, through Internet companies, or from mail-order companies. Shop around for good prices and sales. Caution: it's easy to overdo it when it comes to shopping for hamsters. There are so many fun products on the market. It's wise to make yourself a shopping list before you enter a store. That way, you'll stick to your list, get everything you need, and stay away from what you don't need.

Did you know?

KEEP YOUR HAMSTER'S cage out of direct sunlight and away from direct heating sources. A temperature range of 65 to 80 degrees Fahrenheit is ideal for hamsters.

Cages with plastic tunnels can be difficult to clean.

Hamster Habitats

A proper cage and all the trimmings go a long way toward keeping your hamster healthy and comfortable. But what type of housing is appropriate? What's inappropriate? Where can you buy cages and cage accessories?

In the world of hamster gear, there are three types of cages: wire, glass, and plastic. All three types are used or advocated by hamster enthusiasts. Cages can be purchased at your local pet supply store, from pet supply catalogs, or online at Web sites.

Even though cages come in a variety of types, shapes, and sizes, your primary concern when choosing housing for your hamster is his comfort and safety. Just as you want to live in comfortable surroundings, so does your hamster. The cage should be spacious, allowing plenty of room for sleeping, playing, and eating. A minimum size of 19 square inches is sufficient, but larger is always better.

Regardless of what type of housing you choose, it must be safe for your hamster. It should be well constructed with no cracks, loose bars, or sharp edges. In spite of their popularity, cages with more than one story and lofts should be avoided. The hamster is burrower, not a climber. Even a fall of six to eight inches could cause him great harm.

One choice in hamster housing is a wire cage. Wire cages are usually the least expensive of all cage types, and they are very durable and chew resistant (yes, your hamster will attempt to chew his home). Enthusiasts who prefer wire housing tout the fact that wire cages are well ventilated, and fresh air helps prevent respiratory problems in these small creatures. Look for a wire cage with a solid, removable bottom rather than a wire bottom. Years ago, wire bottom cages were popular, but these can be hard on the hamster's feet, causing sores. To prevent break-

Glass aquarium tanks make a comfortable home for hamsters.

outs, look for a cage that has a bottom that latches to the walls and a door that latches securely. Horizontal cage bars should be about $1/2$ inch apart for golden hamsters; dwarf hamsters may need caging with at least $1/4$ inch bar spacing to prevent escapes.

Glass aquarium tanks make a comfortable home for hamsters (some pet supply stores sell a plastic version), and they are usually inexpensive. The tank is easy to clean, and if fitted properly with a metal and screen top, is escape proof. The aquarium should be at least a 10-gallon size for a single golden hamster, but larger is always better. One drawback to aquarium tanks is they can have poor ventilation. That's why a screen top is essential. The lid should fit tightly, but some owners place a heavy object on top as extra security against breakouts. A water bottle adapter may be necessary, as well.

By far the most eye-appealing cages to owners are commercially made plastic tunnels and compartments, which can be assembled to simulate the hamster's wild habitat of tunnels and burrows. These brightly colored enclosures give the hamster an opportunity to scurry and run through a maze of tunnels and compartments. Hamsters seem to enjoy them tremendously. This type of housing does have a few downsides: it's more expensive than other kinds, it's difficult to clean (a special brush is required for the tunnels), and it's chewable (leading to possible escape and digestive upset). If you decide to buy one of these cages, be sure to buy a large model. Beware of models marketed as "starter" homes. They're usually much too small.

Some owners buy plastic tunnels along with a wire cage or glass aquarium. The hamster is kept primarily in the cage for eating and sleeping and then placed in the tunnels for exercise. This is a great way to safely expand your hamster's world. He is able to run about as he would naturally but is protected from dangers present in your home.

In the early days of keeping hamsters, fanciers often built their own cages out of necessity. Hamster products as we know them today weren't available. Owners had to make housing for their hamsters, often from wood and wire, similar to guinea pig or rabbit housing. Perhaps you'd like to make a cage for your hamster, to save money or because you're crafty. But building your own caging isn't necessary or even recommended today. Housing isn't that expensive. If buying a cage stretches your budget too severely, it's probably best to delay owning a hamster until you're financially stable. Manufactured housing is made specifically with hamsters in mind. It's safe, stable, and made of appropriate materials. Wood, for example, of which early cages were made, is chewable and difficult to clean. It is easily soaked with water and urine. It's best to leave cage making to the experts and buy one. Put your crafty talents to work making furniture instead!

Housing for Dwarfs

Housing a dwarf hamster is slightly different than housing a golden hamster. Goldens are loners, whereas dwarfs are more sociable and should be kept in pairs. Dwarfs mate for life and crave contact with their fellow species. However, if one of the pair dies, it is best left alone. Introducing a new dwarf isn't usually successful.

The dwarf hamster's diminutive size is adorable, but it also enables the small creature to slip through cage bars that keep a larger golden hamster enclosed. Dwarfs are best kept in glass or plastic aquarium tanks with tight-fitting lids or in wire mouse cages that have bars with very close spacing.

Accessories

Regardless of the type of housing you buy, the basic furnishings are the same: water bottle, food bowl, nest box, bedding, exercise wheel, and toys. Depending upon what kind of housing you buy, you may need to purchase a water bottle (some housing comes packaged with a water bottle). Be sure to buy a plastic and steel water bottle, not rubber and glass. The glass could break and cause great harm to your hamster. Most owners use a medium-size gravity water bottle, the kind that hangs on the outside of the cage with a metal ball in the tip to regulate flow. Some enthusiasts like to offer water in a dish, but this can be a messy idea. Clean, fresh water is essential for your hamster's

Most hamsters like to climb in their food bowls to eat.

Hamsters love to burrow in their bedding.

health, so fill the water bottle daily, and wash the bottle with soap at least once a week.

Make sure food dishes are on your shopping list. Because hamsters like to climb into their bowls to eat, enthusiasts recommend heavy ceramic crocks made especially for small animals. Lightweight dishes will be knocked over and chewable dishes will be chewed. Buy a crock that's not too big or deep but can hold enough food for a day. You may find the food bowl empty because the hamster has hidden the food somewhere in the bedding. This is normal. Simply fill the bowl daily with healthy foods.

You want to give your natural burrower a place to hide by buying a nest box. A nest box is a small cubbyhole with bedding where the hamster sleeps, hides, or plays. A nest box gives the hamster a sense of security. It can be made of wood, ceramic, plastic, or woven grass, and there are many different styles on the market. You can also use a small cardboard box (which will be chewed) or buy one made specifically for hamsters (some are chew proof). Place the bedding inside the nest box so the hamster can make a comfortable resting place. Depending upon the size of the cage, you may want to buy more than one to give your hamster plenty of hiding places.

Bedding is a must for your shopping list. The hamster's home must be "carpeted" with safe, dry bedding, which is easily purchased at pet supply stores. Hamsters like to burrow and enjoy digging through the bedding to take a nap, hide food, or, for new mothers, hide babies. Some good bedding choices include wood shavings, recycled paper, paper pellets, or corncob bedding. Aspen shavings are good. Avoid cedar and pine shavings, because these woods contain oils that can be toxic to hamsters. Buy only bedding made specifically for hamsters, and do not use cat litters.

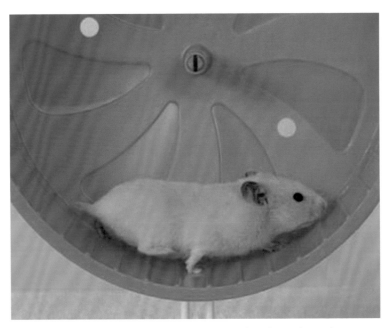

A solid-bottom wheel is better for a hamster's feet than a wire one is.

Exercise wheels are a must for hamsters. Most exercise wheels are made of wire, but there are some with solid bottoms. Many hamster enthusiasts recommend those with a solid bottom rather than wire for safety reasons—the hamster's small feet can be caught in wire wheels, as can the hair on a longhaired breed. The diameter of the wheel must be at least 5 in. to be safe for a golden hamster. Dwarfs can use a mouse wheel. The wheel may squeak, but that's easy to remedy—just apply a drop of mineral or baby oil to the axle.

Hamsters are playful creatures, so toys are a must. Toys are not only fun, but they also have the serious purpose of providing healthy stimulation. Be careful to choose only safe toys for your hamster. They should be the correct size for your hamster, with no sharp edges and in good working condition. There are many different kinds of toys on the market, some for hiding, and others for cruising. You can purchase manufactured toys or make some yourself. A very popular toy is a hamster ball, a hollow plastic ball in which the hamster is placed so he can run safely enclosed in the ball. A variation of this is a hamster car. You can also give your hamster simple items found at home to play with, such as empty toilet paper rolls or small cardboard boxes. Of course, these don't last long with your hamster's chewing capacity and need to be replaced often. Manufactured toys tend to last longer.

Hamster chow is another item to add to the list. Purchase a commercially prepared hamster mix at the local pet supply store. Then fill the bowl, and dinner is served. You can add variety to your hamster's diet by serving fruit: apple or pear in small portions. Carrots, spinach, celery, cauliflower, or corn are good vegetable choices (see Chapter 7 for in-depth feeding information). Some enthusiasts recommend snack sticks, which are sweetened seeds stuck together. Hamsters love to eat them, and they help wear down the hamster's ever-growing rodent teeth.

You need a safe way to carry your hamster home from the pet store or breeder, and off to the vet or show ring. A carrier is also a good place to put your hamster while cleaning his primary cage. Small, hard plastic animal carriers work well for these outings.

If you're purchasing a hamster with long hair, you will need to comb him once or twice a week. Purchase a hamster comb to help keep the coat clean. A toothbrush also works well.

Decorating

How should you set up your hamster's habitat? Carefully, just as you might decorate your own home for comfort and safety. Buy the biggest cage you can afford so your hamster has plenty of space.

Resist the temptation to fill up the entire cage with accessories. You don't want to overcrowd the cage. A wheel, nest box, food bowl, water bottle, and two toys are enough. You want to leave plenty of space for moving around. To keep your hamster interested in his toys, change them every few days. Just as young children become disinterested in a toy they play with daily, your hamster can also become disinterested in his toys. You also need an attachment to hang the water bottle securely. Don't place it too high or too low. You may wish to place a small bowl under the bottle to catch drips.

Toys provide healthy stimulation for a hamster.

CHAPTER FIVE

Hamster Homecoming

CARING FOR A PET IS A FAMILY EFFORT. REGARDLESS OF who's the primary owner of the hamster, all family members, adults and children, should know how to care for her properly. That's why it's helpful to call a family meeting and talk about the new pet. You will need to discuss the dos and don'ts of care and feeding, make sure everyone knows how to handle the hamster properly, and delegate hamster chores. If the hamster is primarily a child's pet—they are wonderful pets for kids—parents must take ultimate responsibility for the hamster's care. Children are very capable of caring for pets, but they need supervision.

Some families have dogs, cats, ferrets, snakes, or birds. How will a pet hamster fit into such a menagerie? Very carefully.

Because hamsters are easily stressed prey animals, the less contact they have with large, potential predators the better. The hamster should remain safely in her cage when you're not holding her. You may think your cat, dog, or ferret is the most docile animal in the world, but you never know what natural instincts will awaken once they're tempted with a small prey animal. Keep other pets away from the hamster's cage so the hamster isn't stressed by their presence.

Veterinary Assistance

An important task in preparing for your hamster's homecoming is to locate a veterinarian in the area who is skilled at treating hamsters. Hamsters tend to be hardy pets, though they certainly

A small-animal veterinarian treats hamsters.

> ## Did you know?
>
> **IT ISN'T SAFE TO ALLOW** your hamster to roam free in the house or even in one room. However, she can enjoy time out of her cage while being handled gently by her owners.

don't live long. You probably won't need to contact a veterinarian in the two to three years of your hamster's life, but the occasion may arise. Finding a competent veterinarian before you need help is essential. In case of an emergency, you will have someone skilled at treating your small pet—and not just any veterinarian is able to do that. You need a specialist.

While there is no formal specialty in veterinary medicine for small mammals such as hamsters, veterinarians specialize in caring for them. Locate a vet who has experience treating exotics. A good way to locate a veterinarian is to ask a hamster breeder or pet supply store for a referral. Involved breeders are usually aware of which vets in town treat hamsters. Or, ask another hamster owner for a recommendation. Additionally, you can contact a local hamster club, 4-H chapter, or the House Rabbit Society in your area (see appendix on page 111). Many vets who treat rabbits are also skilled at treating other small mammals. Look in your local yellow pages for veterinarians who are listed as having an "exotic" practice. The Internet is another good place to search for a veterinarian. You can check out hamster Web sites and chat with other hamster enthusiasts (see appendix on page 111).

If you cannot locate an exotic animal specialist in your area, choose a local vet who is willing to consult with an exotic animal specialist in another city. A good doctor will assume hamsters require special care and will be willing to consult experts

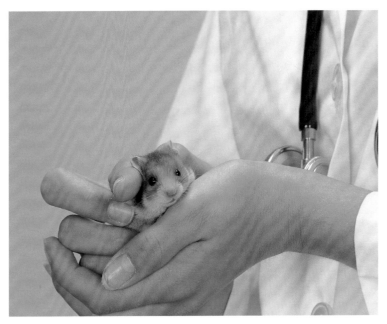

Ask a breeder for a referral to a veterinarian who specializes in exotics.

to find out what that care entails. You can also contact the Association of Exotic Mammal Veterinarians (see appendix on page 111) to find an exotic animal vet in your area.

When you locate a vet who you think can care for your hamster, ask him or her a few questions. How many hamsters and small mammals does he treat each month? What is his medical training in regard to pocket pets? Is he available for after-hours emergency care? One reason it is important to know a veterinarian familiar with and skilled at treating hamsters is the animals' sensitivity to certain medications. Hamsters can have severe allergic reactions to certain antibiotics, including amoxicillin, ampicillin, streptomycin, tetracycline, and gentamycin. This is not a complete list, but a knowledgeable veterinarian knows precisely which medications are safe for hamsters.

Safety Thoughts

Once you bring home your hamster, you are responsible for her safety. It's good to think ahead about how you can provide your pocket pet with a safe home. First, make sure the cage is well made, secure and free of sharp edges that could harm the hamster. Place the cage where the hamster is safe from unwanted visits from other pets in the house. Even if the hamster remains in her cage, the sight of a large dog panting and staring at her from the outside is very stressful. Be sure there are no electrical cords within reach of the cage that the hamster could chew, and do not place her near strong-smelling cleaning products or paints. These could be harmful to the hamster's respiratory system. Do not place the cage in direct sunlight or where there are drafts. Many enthusiasts recommend placing the cage on top of a table.

Unfortunately, hamsters are quite good at escaping from their quarters, especially the very small dwarf hamsters. This is usually due to human error, such as a door left ajar or a lid not securely fastened. Be diligent about keeping the hamster's cage

Naming Your Hamster

ABBY? OREO? EARL GRAY OR ELMO? HOW ABOUT Hamlet or Honeydew? An important part of welcoming your hamster into your home is choosing a name—the right name for your individual pet. This isn't as easy as it sounds. You may need to give it some thought, or you may need some help. A great way to get name ideas is to search the Internet. Type "hamster names" in your Internet service provider's search engine to find an extensive listing of sites that offer help naming hamsters. Have fun and good luck!

Keep the cage out of direct sunlight.

in good repair and shutting doors carefully. Keep on your toes! If your hamster escapes, there are many dangers that await her in the house, such as a dog or cat, poisonous plants, or heating duct vents.

Check-in and Arrival

The day you bring home your pet hamster is exciting, especially for children. To make it an easy transition for the hamster, carry home the hamster in a small travel carrier lined with a little bedding from her former home at the pet store or breeder's home. The familiar scent helps the hamster relax on her adventure into the unknown.

Prior to picking up the hamster, set up the cage with the amenities. Line the cage with dry bedding, fill up the water bottle, place food in the dish, and have toys placed throughout the cage.

Hamsters love to chew.

The Bottom Line

OWNING HAMSTERS IS NOT A TERRIBLY EXPENSIVE hobby, but it does require some funds. You will need to take that into consideration when choosing and bringing home a hamster pet. Initially, you will need to spend a good sum to purchase everything you need for proper hamster care such as a cage, water bottle, dishes, hamster care book, and food. Expect to spend approximately $100 to get started. Caging, especially if you buy setups with plastic tunnels, is your most costly purchase. Hamsters are fairly healthy, so veterinary costs are usually minimal.

Make sure the cage is in a quiet spot in the home that won't be disturbed by other family pets. The next step is to place the hamster in her new home. If the cage is large, place the travel box inside the cage and open the door. Let the hamster venture out when she is ready to explore. Once she comes out, empty the bedding from the carrier box into the cage and remove the carrier.

Resist the urge to handle and play with the hamster right away. Allow the hamster 24 hours to acclimate to her new surroundings. Remember, moving isn't easy! Leaving the hamster alone to adjust is really difficult for kids, who are eager to play with their new furry friend. Don't dismiss the children, but ask them to sit quietly next to the cage and watch the hamster. That way, they can feel included but also can respect the small hamster's need to become accustomed to a strange environment.

CHAPTER SIX

Life with a Hamster

Y OU PROBABLY ALREADY HAVE AN IDEA OF WHAT IT'S going to be like to own a hamster. You can imagine feeding him a favorite treat, watching him run on the wheel, holding him, and showing him off to a friend. But there's much more. The following is a rough sketch of what you can expect.

Night Owl

When you decided to bring a hamster into your home, did you realize you were taking in a night owl? Hamsters are nocturnal, which means they sleep during the day, but come sundown, they come alive. Hamsters sometimes wake up during the day for short periods of time, but they are most lively at night. The best time to play with your hamster is early evening, when he is fully awake and ready for fun.

Because of these natural sleep rhythms, you must carefully consider where in your home you place your hamster's cage. Unless you're a night owl, too, your hamster's nighttime activities may be disturbing while you're trying to sleep. Place the cage away from your bedroom so the hamster can go about his normal behaviors at night—playing—and you can go about yours—sleeping. That's the perfect compromise for both you and your hamster.

Why Did My Hamster Do That?

After observing your small pet for a while, you're sure to notice interesting behaviors. For example, sometimes your hamster stands on his hind legs and sniffs or lays on his back very still. What do these behaviors mean? Physical behaviors are the outward expression of what your hamster is experiencing internally. Some call this body language. Like any language, it requires study to understand its meaning. You can understand what your hamster is feeling by simply watching him and interpreting his physical actions. Look for these behaviors.

- A content hamster can be seen stretching and yawning with half-closed eyes, or washing his face and ears.
- Reciprocal cleaning between mother and babies and between females and males shows contentment and affection. Stroking the head with a paw can denote tenderness.
- A frightened hamster may lie on his back very still or sit up on his hind legs and take a whiff. Continual face washing can also signify fear.
- A hamster on the defense, such as a male attacked by a female, can be seen raising both front paws. This is believed to delay or prevent a more serious attack.
- A youngster who is showing submission and fear to an

This hamster is yawning.

adult might walk stiff-legged, with his tail stretched out and his hindquarters turned toward the older hamster.

• An interested or excited hamster may sit up, or even pop up in the air.

Check the Thermostat

For hamsters, the temperature needs to be just right—65 to 80° F is optimum. You must place the hamster's cage in a place where the temperature is within this range and where it is free from direct sunlight or drafts. The best place for the cage is inside the house, not outside or in the garage, where you can't control the temperature.

If exposed to low temperatures, the naturally desert-dwelling golden hamster will fall into what appears to be a state of hibernation. It is not a true hibernation but a dormant sleep.

Quiet Pets

HAMSTERS ARE VERY QUIET PETS, AND THE MOST common sound owners hear from their cages are squeaks from the exercise wheel. But hamsters do vocalize at times, so if you're very observant, you may be fortunate to hear your small friend expressing himself verbally. Not just wheels squeak, but hamsters themselves squeak. This is believed to be a mating call or a sound the hamster makes when excited or afraid. Hamsters scream when they are distressed or surprised. Like you, hamsters cough or sneeze when they are sick with a respiratory ailment. When annoyed or alarmed, hamsters grind their teeth.

The hamster curls up and is in a deep sleep that is sometimes mistaken for being dead.

The dormant hamster should not be allowed to remain in this deep sleep. He may not be strong enough to recover. He should be awakened slowly by gradually warming up the environment. Once the hamster is fully awake, take care to prevent this from happening again by keeping the hamster where he isn't exposed to temperature extremes.

Handle With Care

The best pet hamsters are those who have been handled, tamed, and socialized from a young age. This requires frequent, gentle contact to teach the hamster to be unafraid of and comfortable with people. Even though some hamsters grow up with little contact from people, there is no such thing as an untamable hamster. An untamed hamster is more likely to be prone to biting.

To make the best of your friendship with your new hamster, begin by letting him quietly adjust to his new home. Don't handle him or bother him for a few days. Assuming you've given your small pet a name, begin using his name and talking with him quietly. Let the hamster get used to you by placing your hand in the cage and gently petting him. Let the hamster sniff you to become familiar with your smell. Offer a treat—sunflower seeds are a favorite—while the hamster is inside the cage. Allow the hamster to crawl on your hand if he wants. Keep still so you won't scare him. Eventually, the hamster will be comfortable sitting or walking on your hand, especially if he associates this with something pleasant like a treat.

Once the hamster is familiar with your hand, you can lift him out of the cage. Use both hands to form a scoop, one hand around the body and other under the rear. Hold the hamster securely, but not too tightly, and keep him close to your body. Pick up your hamster this way several times a day. Talk softly and be very gentle. A too-tight squeeze could hurt the small animal.

Follow these dos and don'ts to be successful at taming your hamster:

- Do not pick up a sleeping hamster. Even a tame hamster bites if startled. Wait until your hamster is awake before attempting to handle him.

- Do be respectful of your hamster's mood. If you notice that your hamster seems frightened or anxious, back off. There's no rush in taming. Place the hamster back in his cage and try again later.

- Do sit on the floor when initially taming your hamster. Young hamsters can be very skittish, and sitting on the floor prevents a fall. Some enthusiasts recommend sitting in a dry bathtub, because if the hamster wiggles away, he is safely enclosed. Once a hamster is tame, he can crawl

Sit on the floor when initially taming your hamster.

on you. Some hamsters even enjoy sitting in shirt pockets! Others like to explore and run around where you are sitting. Be very careful, though, because hamsters are quick and can dart away from your grasp in an instant.

Cleaning Detail

Life with a hamster includes chores. But because of the fun and entertainment these small pets provide, most owners don't mind the cleanup at all. In fact, they're happy to do it because they know dirty cages are a breeding ground for bacteria and viruses.

First on the cleaning list is soiled bedding. You will notice that the hamster tends to use one place in the cage to urinate and defecate. Remove that soiled bedding every day. Then, once each week, completely remove all bedding and replace it with fresh, clean bedding.

While the cage is empty of bedding and your hamster is safely waiting in an alternate cage or travel carrier, wash the cage, tubes, tunnels, nest boxes, and toys thoroughly with mild soap and water. Hamster enthusiasts advise against using harsh cleaners with strong

Clean your hamster's cage and accessories at least once a week.

smells because they could be irritating to the hamster's respiratory system. Rinse well and let everything dry completely before adding fresh bedding and accessories. Then return your small friend to his habitat. The food dishes and water bottle should be cleaned at least once a week; more often if necessary, especially if serving fresh food, which should be removed daily.

Easy Grooming

When it comes to grooming your hamster, it's a breeze. The hamster does it! Hamsters are clean animals and groom themselves many

Use a toothbrush or hamster comb to gently brush your longhaired hamster.

times throughout their waking hours. You don't need to worry about keeping your hamster clean. He can take care of that on his own. If you own a longhaired hamster, you can gently brush his fur with a hamster comb or toothbrush to remove bedding that may be stuck to his fur.

Bathing isn't recommended for hamsters. It's not necessary because they are naturally clean, and it could chill the animal, resulting in illness. Additionally, hamsters cannot swim and could drown if placed in water over their heads.

Leaving Town?

Everyone needs a holiday now and again. If you're planning a vacation, consider care for your hamster while you're away. Hamster enthusiasts say this easy-to-care-for pet can be left safely at home by himself for two to three days. Clean the cage thoroughly before you go, and fill it with fresh bedding. You also need to fill up the food bowls and leave two bottles of fresh water and some fun toys. Batten down the hatches securely to prevent a break out.

If your plans take you away from home longer than two or three days, ask a trusted neighbor or friend to come into your home to care for your pet. Or hire a bonded pet sitter. (See appendix on page 111 for pet sitter information.)

Escape!

The hamster is an inquisitive creature and thrives on exploration. Unfortunately for owners, that drive can sometimes mean escape. Take every precaution to prevent escape by housing your hamster securely and always closing the cage properly. But there are clever hamsters who find ways to break out in spite of your best attempts to prevent it. If this happens, don't despair. Instead, get busy looking for your lost pet.

First, close all the doors and windows. If you're not sure which room the hamster is hiding in, place bowls of food in each room. The scent of a favorite treat may draw the hamster out of hiding because this small creature has an excellent memory when it comes to food sources. Hamsters have relatively poor eyesight and rely on smells to guide them. If you notice the food disappearing in one room, you know the hamster is there.

Once you know which room your hamster is in, search it. Look everywhere. In many cases, the hamster seeks out a dark cubbyhole to hide in and make a nest. They are especially fond of hiding in stacks of magazines or newspapers, boxes, and even chair cushions. Place the hamster's cage in the room, leaving the cage door open. Your pet may wish to return to his familiar

If your hamster escapes he may chew on electrical cords.

Ouch!

LIKE MANY CREATURES, HAMSTERS BITE IF GIVEN sufficient reason. And it hurts! Most hamster bites occur when the animal is startled or grabbed suddenly. It is best to handle your hamster very gently, scooping him up and cradling him in the palms of your hands. The more a hamster is handled, the tamer he becomes.

Do not eat lunch and then handle your hamster. Hamsters have a strong sense of smell, and the scent of food on your hands may prompt the hamster to take a bite. The hamster is not being hostile, he's merely sampling what he thought to be a snack. Do not scold your hamster. Most biting is the result of fear, not viciousness. If you get bitten, try to think what you did to cause it. The most common reasons for biting include picking up a sleeping hamster, grabbing a hamster from above, and accidentally pinching the hamster when you pick him up.

If you are bitten, calmly place the hamster back in his cage. If the bite is severe and bleeds, disinfect it with iodine scrub or another disinfectant. A thorough washing after a bite is usually enough to prevent an infection, but if it swells severely, consult your physician. Rabies is a possibility with any warm-blooded animal, although it's very unlikely from a hamster. Hamsters are not routinely vaccinated against rabies as are dogs and cats.

surroundings. You can even try placing the hamster's squeaky wheel in the room. If the hamster goes to it for a run, you'll be able to hear it.

Don't give up if you fail to locate your hamster immediately. Stories abound of lost hamsters who turn up days or weeks later in places such as the kitchen pantry or basement. Keep looking. Your small friend is likely to come home.

A hamster might bite you if you pick him up right after you eat lunch.

CHAPTER SEVEN

Dinnertime!

IF YOUR NEW PET HAMSTER COULD SPEAK, SHE WOULD probably ask, "What's for dinner?" Don't be offended by that thought. Hamsters just like to eat, and it's your job as a responsible owner to provide a tasty, nutritious menu. Fortunately, that's not difficult. Feeding hamsters is not complicated, and there are many commercially prepared diets on the market. All you need to understand is the hamster's nutritional requirements and where to purchase the food.

Wild Dining

Did you know hamsters are omnivorous, meaning that in the wild these small critters feed on both animal and plant substances? Because pet hamsters are most frequently seen eating just seeds and vegetables, many people do not realize wild hamsters eat insects, worms, or other animal substances.

No need to imitate a wild diet verbatim for your pet hamster, though. Commercially prepared diets, in conjunction with fresh vegetables and fruits, provide your hamster with appropriate nutrition. Most hamster enthusiasts agree that feeding meat isn't necessary as long as the diet fed contains roughly 16 percent protein. However, adding cooked meat, milk, yogurt, crickets, or mealworms to your hamster's diet once a week is acceptable and a good way to make sure the hamster receives adequate protein.

The Basic Diet

You can feed your hamster daily with what's commonly called a dry mix. You can find a dry mix for hamsters at most any pet supply store, locally or on the Internet. The mix usually contains a mixture of seeds, nuts, oats, corn, or barley. Seed mixes are a favorite with hamsters, but because some hamsters pick and choose their favorite seeds (especially sunflower), they may not receive adequate nutrition by eating only a seed mix. Hamster enthusiasts recommend feeding a pelleted diet in addition to ensure optimum nutrition. Pelleted or block diets are made to meet the hamster's complete nutritional needs. Look for one with at least 16 percent protein. The hard consistency of these diets encourages gnawing, which keeps the hamster's ever-growing teeth at a proper length. Though complete nutritionally, pellet or block diets are bland compared with seed diets, and some hamsters reject them. However, if the pellet diet is introduced at an early age, your hamster will probably eat it. Hamsters generally are not picky eaters, though they do show preferences.

Most pet supply mixes are a favorite with hamsters.

Hamsters may not receive adequate nutrition by eating only a seed mix.

Fed together, a dry mix and pellets provide your hamster with the staple of a good diet. Be sure to monitor your hamster to make sure he isn't picking and choosing certain foods. The idea is to encourage variety, which is the best way to ensure your hamster receives all the nutrients he needs. Some hamster enthusiasts recommend feeding only a small amount of seed mix. That way, the hamster is still hungry and will eat the pellets or block diet. Shop around for your hamster's food. There are many seed and pellet diets on the market, most of them packaged quite colorfully. Do not purchase avian seed diets. Buy a diet made specifically for hamsters, and make sure what you buy is fresh.

Fresh Foods

Many hamster enthusiasts recommend adding fresh foods to the diet with the idea that a wide variety of foods provides needed nutrients. In addition, fresh foods are generally healthier than dry, processed foods. You can safely feed your hamster a small amount of fruits and vegetables in addition to his basic diet. Some hamster favorites include apples, pears, pitted cherries, seedless grapes, carrots, broccoli, celery, peas, and cauliflower.

When feeding fresh foods, don't overdo it—two to three times a week is plenty. If your hamster isn't used to fresh food, start slowly. Don't give him a huge slice of apple. He may eat the whole thing and end up with an upset stomach. Introduce new foods one at a time. Wash all fruits and vegetables thoroughly, and cut them into tiny pieces before serving them to your hamster. Remove and toss uneaten portions daily so they don't spoil.

Treats

Pet owners love to give their pets treats, and hamster owners are no exception. Offering a treat is a way to show love and affection, and it can also be a way to win your hamster's affection. Handing out goodies makes many people feel as though they're caring, loving owners.

While lavishing your hamster with treats may create a feeling of well-being, it can create problems. Offering unhealthy treats or too many treats can cause your hamster to shun her regular diet, which results in nutritional imbalance. In time, a hamster that isn't receiving the proper balance of nutrients, having given up her regular diet for a feast of treats, will become ill. That's why it is imperative to be careful about treats.

Giving your hamster healthy treats once a week is fine. But it is important to recognize that treats are not essential to your

Hamsters are hoarders and are often seen keeping food in their cheeks.

Common Poisonous Items

THE FOLLOWING ITEMS, PLANTS, AND HOUSEHOLD products are potentially dangerous to your hamster if nibbled, ingested, or contacted directly. Chances are you won't need to be concerned about your hamster having access to these items because she'll stay secure in her cage. However, breakouts and mishaps do occur. The following is not intended as a complete list of dangerous food items. If you are concerned about something your hamster has eaten or chewed, contact your veterinarian. You can also contact the ASPCA's Animal Poison Control Center (APCC) at (888) 426-4435 (a consultation fee is billed to the caller's credit card).

- alcohol
- avocado
- bleach
- coffee
- english ivy/yew
- flea/tick products
- holly
- insecticides
- lima bean
- mothballs
- navy bean

- antifreeze
- bird of paradise
- chocolate
- detergents and disinfectants
- eucalyptus
- gasoline
- hyacinth
- lighter fluid
- mistletoe
- nail polish
- oak

- oleander
- oven cleaner
- paint
- poinsettia
- rhubarb
- rhododendron
- soap
- tobacco
- turpentine
- weed killer
- yellow jasmine

Did you know?

HAMSTERS HOARD.
What they don't eat immediately is likely to be hidden away in their nest. Feed fresh food sparingly, otherwise the hamster may hide it away where it may spoil.

hamster's well-being. What is a healthy treat for your hamster? Peanuts, dry toast, cooked pasta, sunflower seeds, breakfast cereal (sugar free), dog biscuits (great for gnawing), raisins, dried cranberries, or steamed rice all make good treats. For added protein, you can offer small amounts of yogurt, scrambled or boiled egg, and cooked chicken (no spices). There are also a variety of commercially made treats available at pet supply stores.

H_2O

Fresh water is essential for your hamster's good health. Change the water in your hamster's water bottle daily. Water is best provided in bottles, preferably with metal sipper tubes. Hamsters may gnaw or break plastic or glass tubes. Liquid vitamins, available at pet supply stores, can be added to the water if necessary. This is helpful if your hamster is a picky eater. A vitamin supplement ensures that hamsters receive all the necessary nutrients. The downside to this is that if your hamster doesn't like the taste of it, she won't drink the water and may then become dehydrated. Before adding a supplement to your hamster's diet, it is wise to consult with your veterinarian. He or she can help you determine whether a supplement is a healthy choice for your pet.

Open Buffet

The healthiest way to feed your hamster is to keep the bowl filled with dry mix or pellets or whatever staple food you choose. The

Hold the Onions, Please

THERE ARE CERTAIN FOODS YOU SHOULD NOT FEED YOUR hamster because they can cause illness or digestive upset. Do not give your hamster these foods:

- any sugary, spicy, or salty foods
- canned or frozen vegetables
- citrus fruits
- iceberg lettuce
- Lima bean
- onion
- rhubarb
- tomatoes
- avocado
- chocolate
- garlic
- kiwi
- navy bean
- raw beans
- sprouting potato buds

- sticky foods such as caramel or peanut butter

Hamsters should have round-the-clock access to food.

hamster's fast metabolism causes it to burn calories quickly. Round-the-clock access to food enables the hamster to eat and replenish as needed. Most hamsters eat primarily at night, given their nocturnal natures. However, many hamsters awaken from slumber during daylight hours, snack, then go back to sleep. Offer fresh veggies and fruits in the early evening when the hamster is fully awake and ready to eat. That way, the fresh foods are eaten quickly and won't spoil.

A Good Chew

Like all rodents, hamsters need a good chew. It's important to provide your hamster with safe items to chew to keep their ever-growing rodent teeth the correct length. Fruit tree branches, wood blocks, and seed blocks are good chew choices and are available at pet supply stores. Not only does the hamster enjoy nibbling on these, but her teeth stay in good shape as a result. If you fail to provide your hamster with safe chewing items, she will find something perhaps not so safe in her environment to chew. It's best to buy chew items made specifically for hamsters. Give only wood and branches that are clean, untreated, and nontoxic.

Pellets can be bland compared with seed diets.

CHAPTER EIGHT

The Healthy Hamster

An important part of caring for your hamster is making sure he enjoys good health. While you cannot control all matters pertaining to your hamster's health, there are many steps you can take to keep your hamster bright-eyed, active, and playful. Some of these steps include practicing preventive care by feeding your hamster a proper diet, being aware of the signs of illness and common ailments, and making contact with a veterinarian who is skilled at treating hamsters.

Preventive Care

Preventive care is providing good care before a health condition arises. It's an effective way to keep your hamster healthy. Let's start with your hamster's diet. Feed the right foods! Remember the saying, "You are what you eat." Eat healthy foods and you'll enjoy good health. Eat junk and you'll pay with ill health. The same is true for your hamster.

Feed your hamster a quality dry mix, some fresh foods, an occasional treat, and fresh water. Change or add foods gradually to your hamster's diet, and feed him consistently. Keep track of what your hamster eats, and if he stops eating or drinking, contact your veterinarian. Chances are, if your hamster receives the proper nutrients from his diet, he is more likely to remain healthy and more able to fight off illness.

House your hamster in a clean, spacious, and safe home. If you house your hamster properly, he is less likely to be stressed. The less stressed your hamster is, the better he is able to fight off illness because stress can impair the immune system and make him more susceptible to illness.

Feed your hamster an occasional treat.

Clean your hamster's cage regularly. A clean environment helps reduce bacteria and viruses. Remove wet or fecal-laden bedding right away. Frequently wash the feed bowls and water bottle. Wash your hands before and after handling your hamster.

Be careful how you hold and handle your hamster. It's important to learn the proper technique so you never accidentally drop your small friend. Even a short fall is traumatic. Teach your children how to hold the hamster, and always supervise children when they are handling the hamster.

While you are handling your hamster, take note of his physical appearance. Be on the lookout for changes such as a wet rear end, hair loss, or weight loss. When you have your hamster out of the cage, supervise him carefully. Don't be tempted to let him roam free. Keep other household pets away from the hamster. If you're alert and watchful, you can stop trouble before it occurs.

Common Health Concerns

Hamsters can suffer from a variety of illnesses—some mild and some serious. If you notice any signs of illness in your hamster, consult your veterinarian right away. Some of the signs to look for include sneezing, discharge from eyes and ears, ruffled coat, loss of appetite, hair loss, wetness around the tail, diarrhea, or inactivity.

Did you know your pet hamster could catch a cold from you? If your hamster catches a cold, you'll notice a runny nose and eyes, sneezing, and lethargy. Most colds clear up in a few days, but to make your hamster comfortable and to keep the cold from progressing into pneumonia, place the cage in a warm area, free from drafts. You can place a light near the cage to add warmth. To prevent colds, wash your hands before and after holding your pet. If you have a cold, avoid handling your hamster.

A condition called wet tail commonly affects hamsters and is potentially serious. This bacterial illness, *proliferative ileitis*, can cause severe diarrhea and can be fatal. Wet tail is highly contagious and most common in recently weaned hamsters who have gone to new homes. The exact cause is uncertain, but the condition is associated with stress, crowding, and diet changes. Affected hamsters may die quickly, exhibiting signs such as diarrhea (causing wetness around the tail), rectal bleeding or rectal prolapse, loss of appetite, and lethargy. Owners must watch carefully for this condition in newly acquired hamsters. Sick hamsters must be quarantined and treated by a veterinarian. Treatment usually consists of antibiotics, fluids, and medication to stop the diarrhea.

It's wise to ask the person from whom you're acquiring your hamster if any of the hamsters have recently been treated for wet tail. Hamsters with this condition must be housed separately and their quarters thoroughly sanitized to prevent the spread of the disease. Even if the breeder reports no incidence of the disease, owners should watch new pets carefully for the first few weeks for signs of the disease.

Diarrhea can be caused by wet tail, but it can also be caused by other illnesses or by a change in diet. Overfeeding of fresh fruits and vegetables is a common culprit. If this is the case, withhold fresh foods for a few days and resume when the diarrhea clears up. Start fresh foods slowly. If you're not sure what is causing the diarrhea, consult your veterinarian.

Abscesses are pockets of infection that form from a cut or scrape to the skin. They can also form in the cheek pouches if the hamster eats abrasive food. Pus accumulates and causes a lump, sometimes noticeable to the eye and other times detectable by stroking the hamster. Abscesses sometimes open and drain on

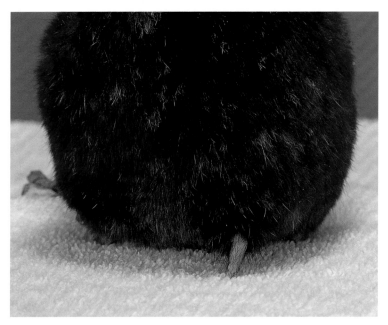

Check your hamster's tail for signs of wet tail.

their own, but most require veterinary attention for draining, flushing, and treatment with antibiotics.

If you notice bald spots on your hamster, he could be infested with mites. In this case, hair loss is caused by incessant itching and scratching. Demodex mites, causing demodectic mange, most commonly affect hamsters. Veterinary treatment is essential.

Hamsters can suffer cuts and lacerations and usually heal well on their own. But a serious cut or bite wound (from a fight with another hamster) may need veterinary attention to heal correctly and prevent infection or abscess.

Not all occurrences of sneezing and runny eyes are due to a respiratory illness. Your hamster could actually be allergic to something, especially if he also has dry skin, hair loss, or reddish feet and acts normal. Hamsters can suffer from various allergies, including reactions to food, bedding, or agents in the environment (such as smoke or cleaning products). Pinpointing what your hamster is allergic to takes some detective work. For example, if you suspect your hamster is reacting to the bedding, remove it and try something different. If the allergic response continues, you know it's not bedding. Because it can be difficult to figure out what is causing an allergic reaction, it's helpful to consult your veterinarian.

Hamsters can have allergies.

A temperature range of 65 to 80° F is ideal for hamsters. Temperatures above 80° F can cause heatstroke. Signs of heatstroke include lethargy, rapid breathing, and unresponsiveness. If your hamster shows these signs, you must act quickly to reduce his body temperature and prevent death. Wrap your hamster in a damp, cool towel or pour cool water on him. Do not use cold water. Call your veterinarian for further instructions.

You can prevent heatstroke by keeping your hamster's environment in the proper temperature range of 65 to 80° F. Do not place the cage in direct sunlight or near a heat source, and always keep fresh water available. During hot weather, run the air conditioner in your home or use fans (which will make you both more comfortable).

The average life span of hamsters varies within the species. Golden hamsters live an average of eighteen months to three years. Dwarfs live eighteen months to two years. As your hamster ages, you will probably notice changes including hair loss, reduced activity, and subtle behavior differences. These are normal changes, but you must distinguish them from signs of illness. If you notice something you're not sure about, talk with your veterinarian.

The Question of Breeding

Perhaps you are considering breeding your hamster. Before you begin, it's wise to think about why you want to breed your pet. Experienced hamster enthusiasts are dedicated to producing quality animals. The best reason for breeding is to improve the breed. A common but poor reason for breeding is to make money. Breeding animals is not a moneymaking endeavor. Any serious breeder will attest to this truth. Don't fall into the trap of thinking

Breeding hamsters is not a moneymaking endeavor.

you're going to make some extra cash. Most likely, you're going to end up with more hamsters than you know what to do with.

If your goal, however, is to pursue the hamster fancy seriously and to show your pet, get involved with a reputable hamster club. Find an experienced breeder who can be your mentor. Read and study as much as you can about hamsters. Then consider breeding your hamster.

Introduction to Reproduction

An amazing fact about hamsters is that they can reproduce at near lightning speed. With sexual maturity at about two months of age, estrus every three to four days, and a gestation period of sixteen to eighteen days, it doesn't take a mathematician to figure out how fast a hamster can reproduce. In the wild, golden hamsters are solitary until breeding time when a male and female mate and then go their separate ways. Females raise the young, which can be a litter of one to eighteen.

Hamsters have a gestation period of sixteen to eighteen days.

Do not touch baby hamsters. Their scent can change and confuse their mother.

A hamster breeder follows nature's course. Females and males are housed separately and then placed together when the female is in estrus. After copulation, the male is removed and the female is left alone. In about two weeks, the next generation of hamsters is born, with closed eyes and no hair, weighing a mere $1/8$ ounce. No special care is required of the mother and young. In fact, the less interference, the better. If the mother is stressed or threatened, she sometimes destroys her young. Do not touch baby hamsters. Once you touch them, their scent changes and confuses their mother. Some breeders give mother hamsters added protein after giving birth; otherwise, a regular diet is fine. The young hamsters eventually grow hair and begin scurrying about. They are weaned at three to four weeks of age and should be separated at four to five weeks of age to prevent fighting or mating.

Finding good homes for the baby hamsters is the biggest challenge of breeding. Breeders must take responsibility for the youngsters they allow to come into the world. The best scenario is to have a list of potential owners before the litter is born. If not, breeders must take care to place the hamsters in homes where the animals receive the best of care.

Hamster Facts

Life span: 18 months to 3 years
Weight: Females 95 to 125 grams
Males 85 to 120 grams
Gestation: 16 to 18 days
Litter average: 5 to 10
Weaning: 20 to 25 days
Heart rate: 275 to 480 beats per minute
Respiration: 50 to 100 breaths per minute

CHAPTER NINE

Hamster Fun

LEARNING HOW TO CARE FOR YOUR HAMSTER IS ESSEN-
TIAL, but as important as good care is, it's not the only aspect
of owning hamsters. You need to factor in some fun! As you'll
learn, hamsters—and the people who keep them—are a fun-loving bunch. Did you know there are hamster presidential candi-
dates? Dancing hamsters and hamster television shows? Hamster
poetry? Here's how you can join in the festivities.

Rules to Play By

One of the first rules of having fun with your hamster is to remember her nocturnal nature. You might enjoy playing during the
day, but your hamster doesn't. She may wake occasionally during
daylight hours, but don't expect full participation until late after-

noon. Do not disturb your hamster during the day while she is sleeping. If you do, she could become resistant to handling. Wait until your hamster wakes up on her own to initiate playtime.

Also, never forget your hamster's safety. Before you offer a new toy, check it carefully. Is it the right size? Is it smooth, without rough edges or parts that the hamster could chew off? Keep interactions between your hamster and other pets, as well as interactions between inexperienced hamster handlers and the hamster, to a minimum to reduce stress.

Keeping Active

Do you realize your pet hamster thrives on activity? In the wild, the hamster scurries about most of the night collecting food. Pet hamsters don't need to search far for food, but they do have excess energy to burn and enjoy doing just that. You can fulfill your hamster's need to keep moving by providing a spacious cage furnished with toys, a wheel, and tunnels.

As you learned in Chapter 4, the hamster's cage should be comfortable and safe. It should be spacious, with plenty of room for sleeping, eating, and playing. A minimum size of 19 square inches is sufficient, but bigger is better. A large enclosure gives your hamster plenty of room to stretch her legs and move about. But don't forget the rule of one golden hamster per cage!

The wheel is a classic hamster toy and is a must for every hamster cage. Once the hamster jumps on the wheel, the race begins! Some hamster enthusiasts do not recommend leaving the wheel in the cage all the time, but rotating it with other toys and furnishings that encourage activity. This helps reduce boredom and encourages curiosity. Make sure you buy a wheel made especially for your hamster (dwarfs may need a mouse-size version.)

Share your enjoyment of having a hamster by joining a hamster club.

Some enthusiasts recommend solid-bottom wheels rather than wire for added safety. To remedy a squeaky wheel, place a drop of mineral oil on the axle.

Another favorite and more modern toy is the hamster ball. The hamster is placed inside a hard plastic ball and runs the same way she would on the wheel, enclosed safely within. The difference is that the ball rolls about your living room, whereas the wheel stays stationary. Hamster balls come in a variety of shapes and sizes and are made by several manufacturers. Some roll freely; others are designed to roll on a track. The balls are especially fun for children, who are fascinated by watching a hamster race about the house.

Plastic tunnels are another excellent way to encourage activity. Hamsters enjoy running in them during their waking hours, perhaps much like they do in the wild. Though difficult to clean and not always gnaw proof, tunnels are fun for hamsters. You can purchase them at pet supply stores, in a wide combina-

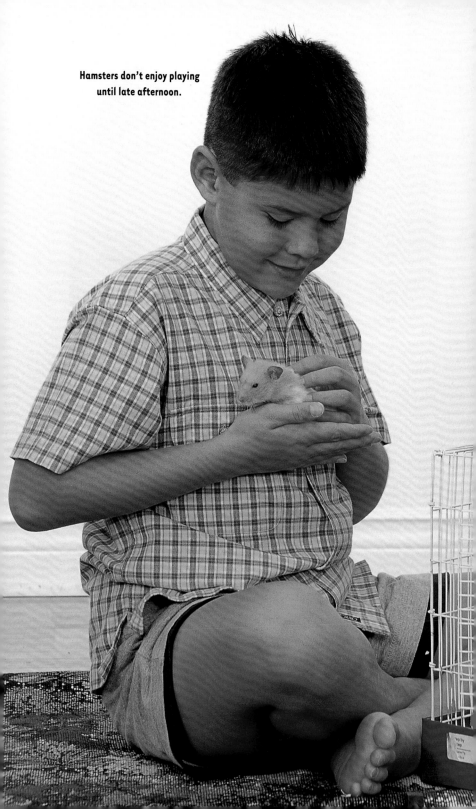

Hamsters don't enjoy playing until late afternoon.

tion of setups. Tunnels are an excellent way to give your hamster time out of his cage safely.

You can also encourage your hamster's activity with simple items found right in your home. Hamsters enjoy playing with just about anything, including paper bags, cardboard boxes, empty paper towel rolls, socks, and clean cloth. Just be sure to place only safe, nontoxic items in the cage, and change toys frequently.

Clubs and Shows

A great way to share your enjoyment of owning hamsters is to get involved with a hamster club. There you can meet others with a similar interest in hamsters. Clubs are also a good way to increase your knowledge about hamsters because members are typically well informed about their small pets. (See appendix on page 111 for a list of hamster clubs.)

Clubs usually sponsor hamster shows, which are competitive events that allow serious enthusiasts to showcase their small animals. Hamster shows can take place on their own, in conjunction with county fairs, or in conjunction with similar exhibitions such as rat and mouse shows.

Hamsters entered in these events are not the average pet hamsters but are specially bred to meet specific criteria and standards of perfection. The standards that breeders seek to meet differ according to breed and include a physical description outlining the desired coat, size, color, and markings. Most hamster clubs in the United States and England have adopted the standards determined by the British Hamster Association, a long-established club.

Like other livestock shows, hamster shows are divided into classes. Entrants are charged a fee, and winners are rewarded with

Show hamsters are bred to meet specific criteria.

Hamsters are judged according to the standards for each class.

rosettes and ribbons. Hamsters are displayed in special show pens, which must be approved by the club sponsoring the show. Hamsters are judged according to the standards for each class.

If you are interested in showing your hamster, begin by joining a well-established hamster club, and ask an experienced enthusiast to be your mentor. Attend as many club meetings and shows as you can. Young hamster owners can get involved by joining a 4-H small animal pet project. For more information on these projects, you can contact your county extension office. When you think you're ready, enter a show. Who knows? You and your hamster might be winners!

The Observation Game

You can learn a lot about your hamster by sitting quietly and watching. For example, eating is a favorite pastime for hamsters, as is collecting and hiding food. Fill up the food bowl, and then watch your hamster as she gathers the goods and fills her pouches. Your hamster will hide bits of food here and there and then perhaps go back for more. Encourage children to watch quietly. It's a great way for them to learn and understand hamster behavior.

Appendix

THERE ARE MANY RESOURCES AVAILABLE TO HAMSTER lovers: books, clubs, and Web sites. The more informed you are, the better you are able to care for your hamster. Check out the following resources to get started.

ORGANIZATIONS

American Fancy Rat and Mouse Association
9230 Sixty-fourth Street
Riverside, CA 92509-5924
Web site: www.afrma.org
AFRMA is a nonprofit international club for anyone who has an interest in rats or mice. They also list information on hamsters.

Association of Exotic Mammal Veterinarians
P.O. Box 396
Weare, NH 03281-0396
Web site: www.aemv.org
This organization was formed by veterinarians to advance the veterinary care of ferrets, rabbits, guinea pigs, and small rodents. Membership is restricted to veterinarians, veterinary students, and veterinary staff only, but they are a great source to find a veterinarian in your area.

California Hamster Association
calhamassoc@hotmail.com
Web site: www.geocities.com/calhamassoc
CHA is a nonprofit hamster club in Southern California

dedicated to educating hamster owners. You can be added to the group's e-mail list by contacting calhamassoc@hotmail.com.

National 4-H Council
7100 Connecticut Ave.
Chevy Chase, MD 20815
(301) 961-2800
Web site: www.fourhcouncil.edu
4-H is the youth education branch of the Cooperative Extension Service, a program of the United States Department of Agriculture. Each state and each county has access to a County Extension office for both youth and adult programs. Check out the branch near you for hamster projects.

The Hamster Society
22 Carr Bank Avenue
Higher Blackley
Manchester M9 8FT
UK
Web site: www.hamsoc.org.uk
The Hamster Society is a British hamster club, though members from throughout the world are welcome. It is affiliated with the British Hamster Association, a coordinating body for the hamster fancy in Great Britain.

The Humane Society of the United States
2100 L Street NW
Washington, DC 20037
(202) 452-1100
Web site: www.hsus.org

This is a national animal protection organization. You can sign up to receive the biweekly e-newsletter, *Pets for Life*, a guide to living with your pet.

National Hamster Council
P.O. Box 154
Rotherham
South Yorkshire, S66 OFL
UK
Web site: www.hamsters-uk.org
The NHC is one of the oldest hamster clubs in the world. It is dedicated to the keeping, showing, and caring of hamsters.

Small Animal Channel
I-5 Publishing, LLC
3 Burroughs
Irvine, CA 92618
Web site: www.SmallAnimalChannel.com
This website is dedicated to caring for small animals such as hamsters, rabbits, guinea pigs, ferrets, chinchillas, hedgehogs, rats, and mice.

Pet Sitters International
201 East King Street
King, N.C. 27021
(336) 983-9222
Web site: www.petsit.com
PSI is a membership organization founded to educate and support professional pet sitters.

BOOKS

Bucsis, Gerry. *Training Your Pet Hamster.* New York: Barron's Educational Series, 2002.

Pinney, Chris, DVM. *The Illustrated Veterinary Guide for Dogs, Cats, Birds, and Exotic Pets.* New York: McGraw-Hill, 2000.

Sikora Siino, Betsy. *An Owner's Guide to a Happy Healthy Pet: The Hamster.* Indianapolis: Howell Book House, 1997.

Sikora Siino, Betsy. *The Essential Hamster.* Indianapolis: Wiley, John & Sons, Incorporated, 2000.

Vanderlip, Sharon L., DVM. *Dwarf Hamsters: Everything About Purchase, Care, Feeding, and Housing.* New York: Barron's Educational Series, 1999.

Von Frisch, Oho. *Hamsters: Everything About Purchase, Care, Nutrition, Breeding, and Training.* New York: Barron's Educational Series, 1998.

Glossary

breed: distinct group of animals descended from common ancestors with similar characteristics, including color, shape, and size

breeder: person who breeds hamsters or other animals

cage: housing for hamsters

dwarf hamster: general term for dwarf breeds

exotic veterinarian: veterinarian who specializes in the treatment of unusual pets such as hamsters, rabbits, ferrets, and sugar gliders

nocturnal: describes animals who are awake at night and sleep during daylight

omnivorous: describes animals who eat plant and animal substances

pocket pets: general term referring to small mammals kept as pets

purebred: pedigreed member of a recognized hamster breed

rodent: any member of the largest group of mammals, *Rodentia*

show: gathering where hamster enthusiasts showcase their prize hamsters

standard of perfection: written description of a specific breed of hamster

variety: subdivision within a breed

wet tail: a bacterial illness that can cause severe diarrhea and is sometimes fatal

Index